RAINBOWS

Thom Klika

RAINBOWS

ST. MARTIN'S PRESS • New York

Design by Dennis Grastorf

Library of Congress Cataloging in Publication Data

Klika, Thom.
    Rainbows.

    1. Klika, Thom. 2. Rainbow in art. I. Title.
N6537.K57A4 1979   709′.2′4   78-21416
ISBN 0-312-66294-7 pbk.

To my loving wife
Mary
for allowing me
to make a spectrum
of myself
and because she
laughs rainbows.

RAINBOWS

A rainbow is an arch of light exhibiting the spectrum colors in their order, due to reflection, refraction and dispersion of light in drops of water falling through the air. It is seen usually at the close of a shower in the quarter heaven away from the sun. In the brightest or primary bow, often the only one seen, the colors are arranged with the red outside. In the perfect rainbow there is another arch concentric with this but above it called the secondary rainbow, in which the color is in reverse order.

Rainbows form when the sun's rays strike millions of rainbows suspended at a $42\,^1/_2$ degree angle from the viewer.

I put the rainbow in the clouds and it shall be a symbol of the world. Whenever I bring clouds over the earth the rainbow will appear in the clouds and I will remember my covenant.

GENESIS: 9

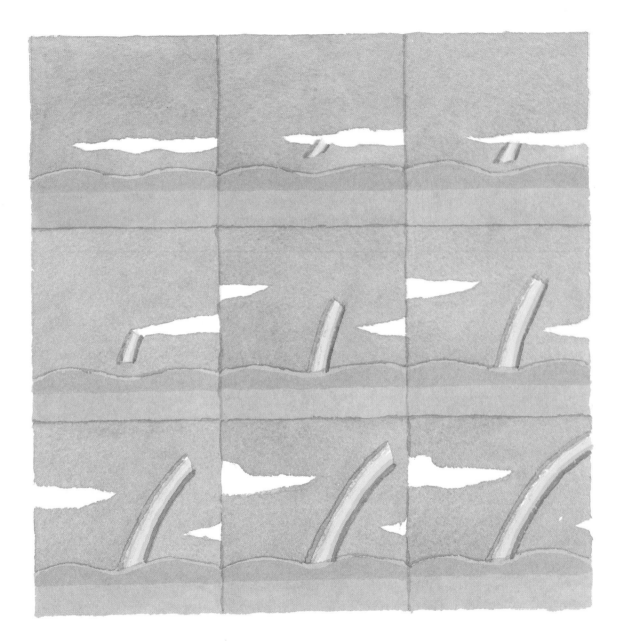

For the love of rainbows . . .
day and night.

To love rainbows
is to respect all creation.

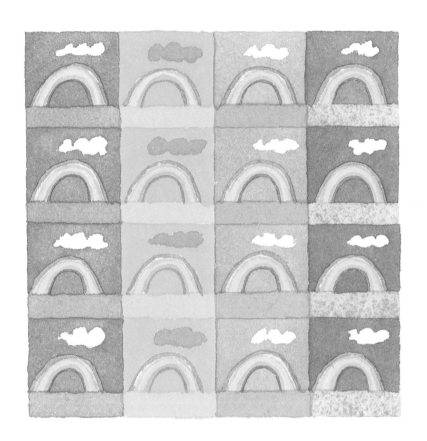

When the tender arm of a sunbeam
embraces the rising mist
of a spring rain,
nature's child is born
. . . . .a rainbow.

USUALLY
WE THINK OF
RAINBOWS AS ARCS
WITH EACH END RESTING
ON THE EARTH'S SURFACE.
BUT THE TRUE SHAPE
OF A RAINBOW IS
A COMPLETE
CIRCLE.

Since the sun must be at the viewer's back, the earth will always obscure half of the rainbow.

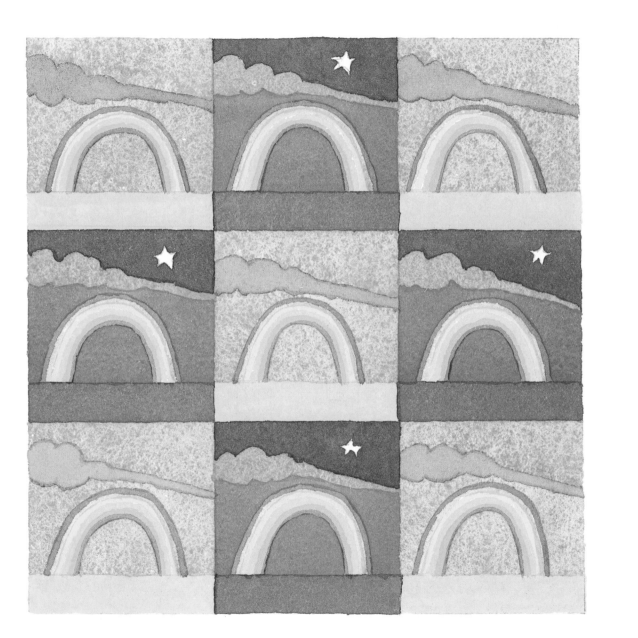

You, whose day it is,
make it beautiful.
Get out your rainbow colors
so it will be beautiful.

SONG TO BRING FAIR WEATHER
FROM THE NODKA INDIANS

Heimdal, the guard, stands on the rainbow bridge that connects the earth to Asgard over this bridge. The gods pass and over it the Valkyrie maidens carry slain Norse heroes to Valhalla.

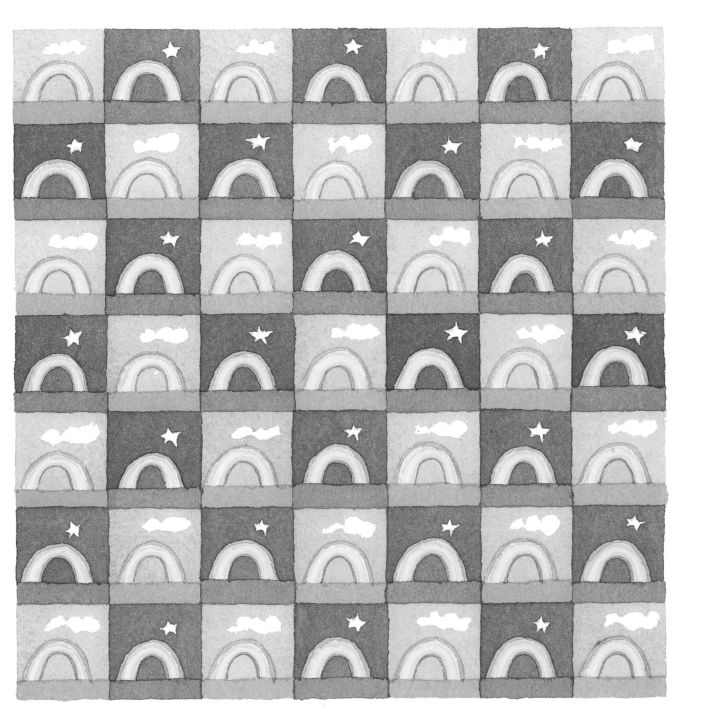

I have journeyed to the
rainbow's end
and have found not gold,
but you, my friend.

—Ian Michaels

Rainbows upside-down;
are like
smiles
rightside-up.

Most rainbows appear
in sunlight, but
they can also be seen
in moonlight.

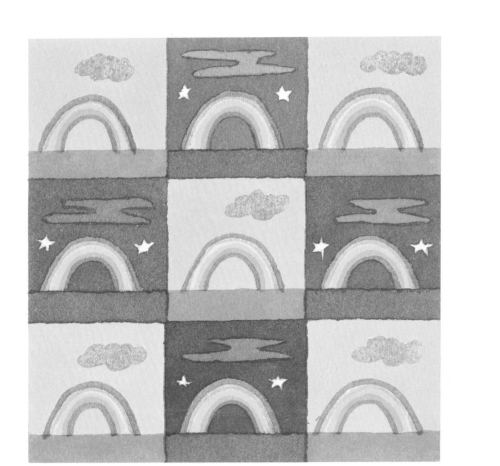

At Cumberland Falls,
Kentucky,
a rainbow
forms over the falls
on nights
with a full moon.

Could
it be
that
rainbows
are
the
iris
of God's
eyes
?

Red and orange,
green and blue,
shiny yellow,
purple, too.
All the colors
that you know
are found
up in
the rainbow.

To a friend's house
　　　　　the road is never long.

There is a calm that flows
into the silence
of the moon's coming,
and through the vivid colors
of a rainbow happening
only the sound
of Om.

A rainbow a day keeps the blues away.

If I cry tears
let them wash away our fears
and make a rainbow of love
for you.

The soul
would
have no
rainbows
had the
eyes
no tears.

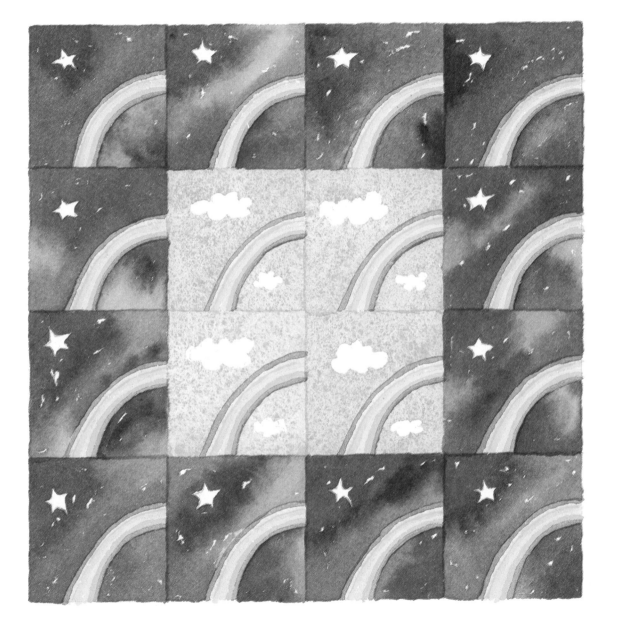

Rainbows keep falling on my head . . .

Rainbows are positive energy.

Rainbow places, objects, and songs: Rainbow Center, Connecticut, Rainbow, Alabama, Rainbow Springs, Florida, Rainbow Chocolate Milk, and a Rainbow Vacuum Cleaner, there's Somewhere Over the Rainbow, Finian's Rainbow, the man who paints rainbows in the sky, there's a Rainbow over the Rio Colorado, the Rainbow Pizza Shop, over 200 Rainbow Motels in the U.S. & Canada, Rainbow Bars & Grills (over 100), the Rainbow Bowling Alley, Warren, Ohio, Rainbow Laundromat in Scranton, Pa., Rainbow Falls, Hawaii. Magic tricks: rainbow ropes and rainbow silks. There's rainbow melons, rainbow citrus and Rainbow Bread. They all go in rainbow delis, diners, and rainbow groceries. There's even a Rainbow Museum, a Rainbow Store, Rainbow Ceramics, Dayton, Ohio. There's a Rainbow Round My Shoulder, I'm Always Chasing Rainbows and Her Beaus Are Only Rainbows, Rainbow Forge, rainbow crafts, rainbow stained glass, Yonkers, N.Y., Rainbow Farm Centers, Rainbow Hangers & Rainbow Lakes, N.J. There's RAINBOW by D. H. Lawrence and RAINBOW STORIES published in 1898. Rainbow makers, rainbow chalk, and rainbow wrapping, rainbow ribbon, rainbow of the Amy Karen Cancer Fund in Beverly Hills, CA. In all, there are some 40,000 rainbows in the U.S. and Canada.

If the Rainbow
came each day do
you suppose we'd
care? Perhaps we
find it beautiful
because it is so
rare!

A
rainbowsaic.

Rainbows occur in summer, fall, winter, and springtime.

Friends are like rainbows.

They bring laughter to your eyes.

Remember:
It takes the sun and
the rain to make a
beautiful rainbow.

An Autobow:
The 1932 Ruxton had a paint
job of rainbow stripes.

Rainbow Bridge National Monument. A monument occupying 160 acres in the Piute Indian Reservation, Utah, was set aside by Congress in 1910 to preserve a natural, arched stone bridge, resembling a rainbow in shape. It is situated 309 feet above a stream bed and has a span of 278 feet.

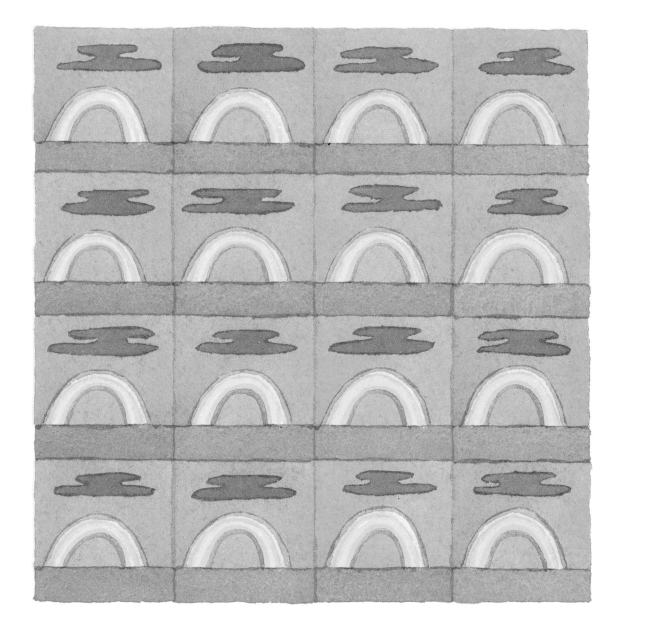

The rainbow and the moon
                    are night lovers.

Iris
was the
Greek goddess
of the rainbow.

The rainbow making machine.

Clouds and Stripes
Forever.